Properties of Matter
Physical and Chemical Changes

by Rebecca L. Johnson

Table of Contents

DEVELOP LANGUAGE

Ice sculptures can be made when the air outside is very cold. A sculptor starts with a block of ice and then carves a design into the ice.

Discuss the photos with questions like these.

What does the ice look like before the sculptor begins the design?

It looks _______.

What do you think will happen if the air surrounding the sculpture warms up?

I think the sculpture will _______.

Describe the ice sculpture in the photo. How is it different from a block of ice?

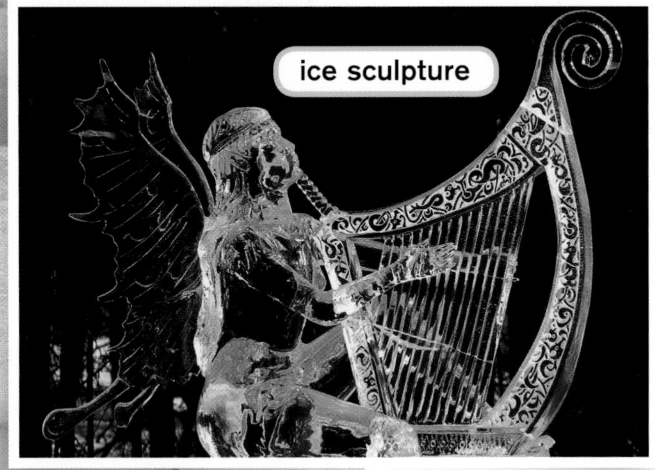

Blocks of ice are cut and taken from here to the World Ice Art Championships in Fairbanks, Alaska.

The ground shakes and there's a loud, rumbling noise. Red-orange lava, or melted rock, shoots out of the top of the volcano. The lava cools to form hard rock.

Lava is a type of **matter**. Matter is anything that takes up space and has **mass**.

One way to describe matter is by its **properties**. Lava is red-orange and very hot. When lava cools, it can form a hard, shiny, black rock called obsidian.

matter – anything that takes up space and has mass
mass – the amount of matter in something
properties – qualities of matter that can be observed or measured

States of Matter

Another property of matter is its **state**, or form. Three states of matter are solid, liquid, and gas. Lava is a liquid. Like other liquids, lava has a definite size but does not have a definite shape.

When lava cools and hardens, it changes state. It changes from a liquid to a solid. A solid has both a definite size and shape.

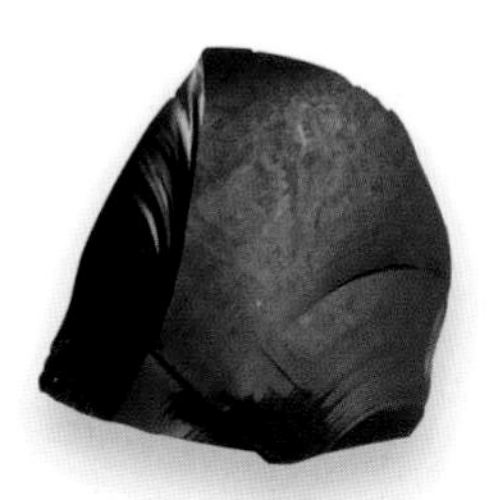

▲ **Obsidian is a solid that can form when liquid lava cools.**

Volcanoes also add gases to the air. A gas has no definite size or shape.

state – the form matter is in

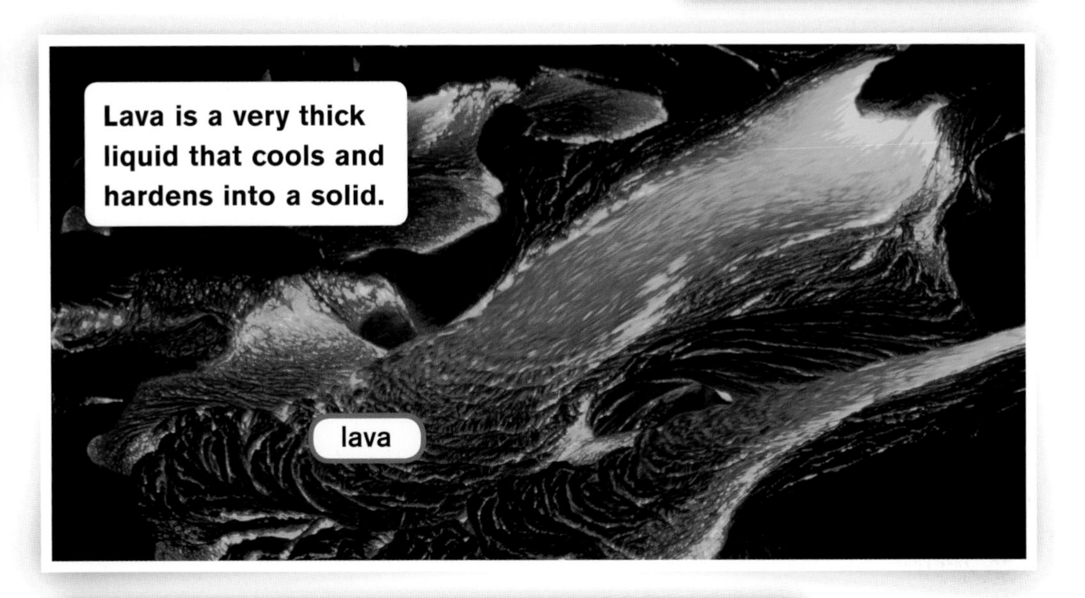

Lava is a very thick liquid that cools and hardens into a solid.

KEY IDEAS Matter is anything that takes up space and has mass. Matter is described by its properties, including its state.

Atoms and Molecules

Solids, liquids, and gases are all made of **atoms**. Atoms are the smallest whole units of matter. But atoms have even smaller parts.

At the center of an atom is its **nucleus**. The nucleus is made up of protons and neutrons. Moving around the nucleus are extremely small electrons.

Two or more atoms can **bond** together to form **molecules**. Most solids, liquids, and gases are made up of atoms that are bonded together as molecules.

atoms – the smallest whole units of matter

nucleus – the central part of an atom

bond – join very tightly

molecules – sets of two or more atoms bonded together

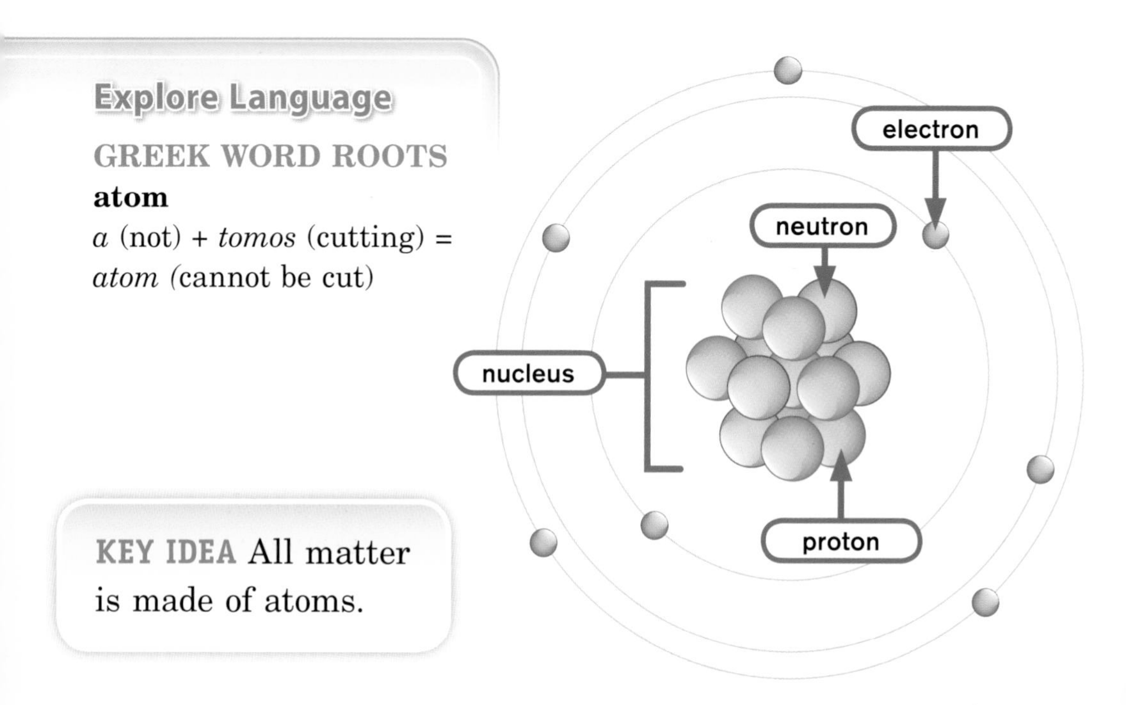

Explore Language

GREEK WORD ROOTS

atom

a (not) + *tomos* (cutting) = *atom* (cannot be cut)

KEY IDEA All matter is made of atoms.

Molecules in States of Matter

Molecules that make up solids, liquids, and gases are always moving. The molecules in a gas are far apart. Because the molecules can move around freely, the size and shape of a gas can change.

The molecules in a liquid are closer together. They still move around, but not as freely as the molecules in a gas. The liquid stays the same size, but it can flow and change shape.

The molecules in a solid are packed tightly together. They cannot move around much at all. The size and shape of a solid stays the same.

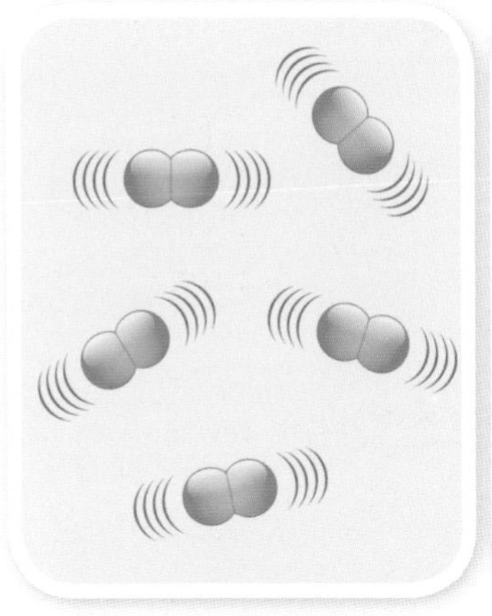

▲ Molecules in a gas are very far apart and move freely.

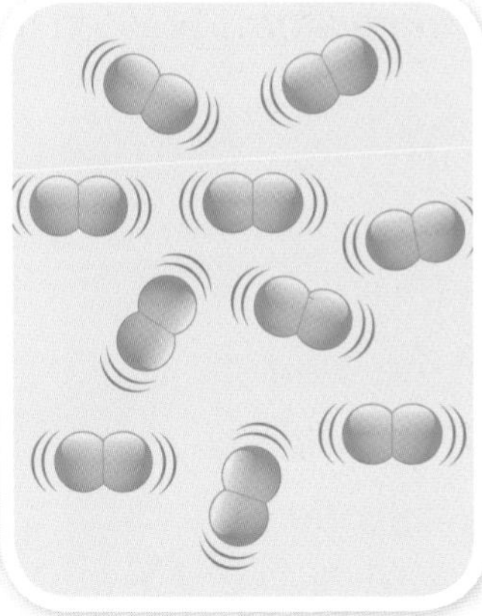

▲ Molecules in a liquid can slide past each other as they move.

▲ Molecules in a solid are moving but stay in the same place.

Measuring Matter

Many properties of matter can be measured. Mass is a property of matter. Mass is the amount of matter in something and is often measured in grams (g).

Length, width, and height are properties of matter. They are often measured in meters (m).

Volume is another property of matter. Volume is the amount of space matter takes up. The volume of a liquid is often measured in liters (L).

The volume of a solid with flat sides can be found by measuring its length, width, and height. To find volume, you multiply the length, width, and height of the solid. The volume of a solid is often measured in meters cubed (m^3).

Volume of a Liquid

▲ **A graduated cylinder measures the volume of a liquid.**

Volume of a Solid

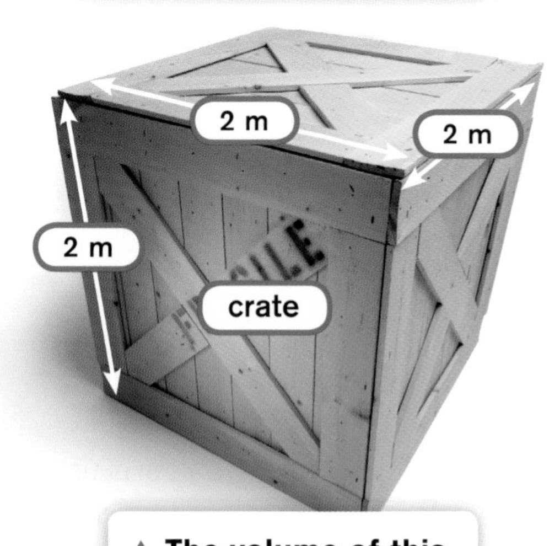

▲ **The volume of this solid is 8 m^3.**

KEY IDEA Many properties of matter can be measured.

INFER

Suppose a frozen juice bar is left out on a kitchen counter on a warm day. After an hour, there is just a pool of juice where the bar had been. Infer what happened to the juice bar.

Did the juice change state? Explain.

In which state did the molecules of the juice move around more freely? Explain.

MAKE CONNECTIONS

In a notebook, list three examples of matter that you can measure. What units would you use to measure each one?

USE THE LANGUAGE OF SCIENCE

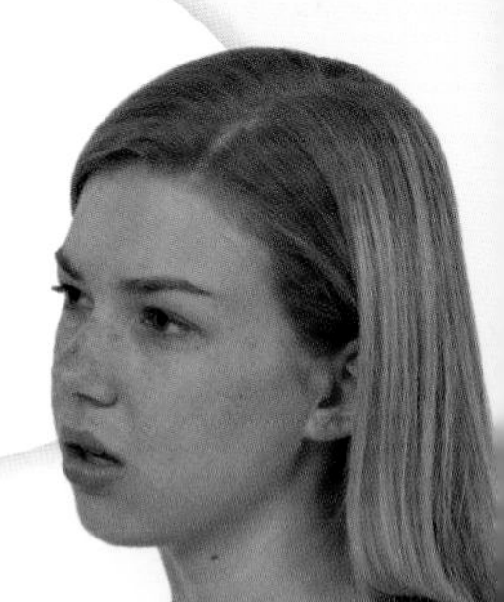

CHAPTER 2

A World of Elements

Gold is a shiny yellow metal. Gold is made of only one kind of atom, so it is an **element**.

All the matter in the world is made of different combinations of elements. Scientists have organized all the known elements into the Periodic Table of the Elements.

element – matter made of only one kind of atom

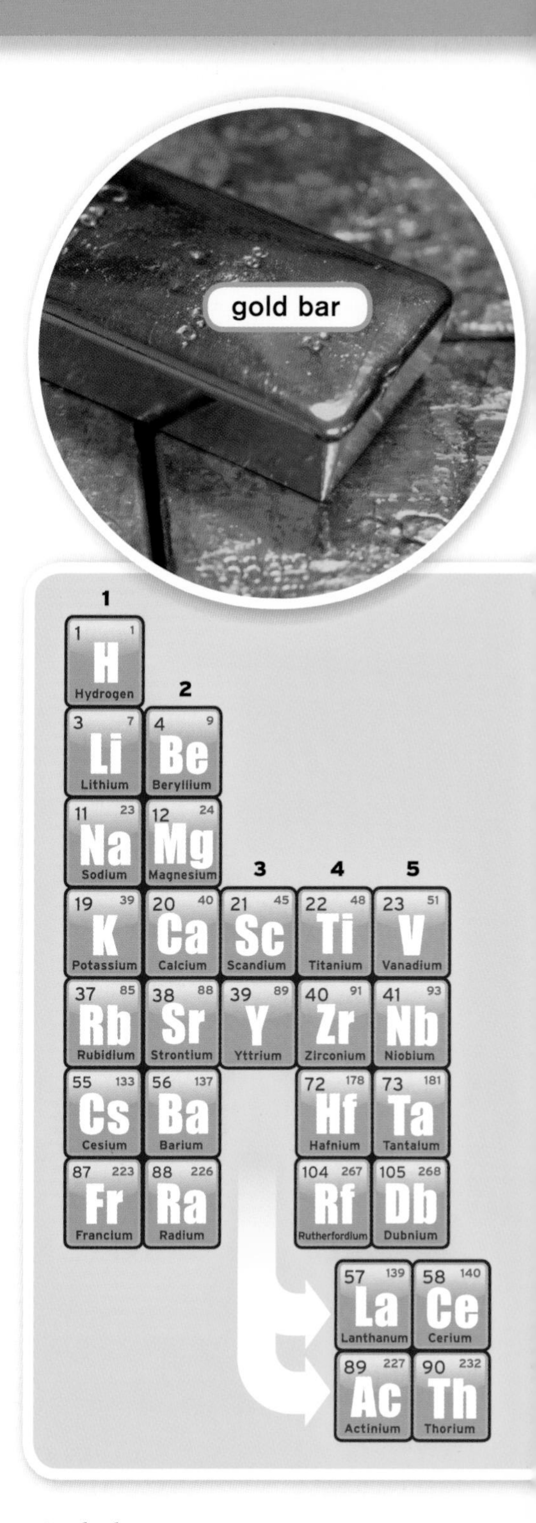

KEY IDEAS An element is made up of only one kind of atom. Elements are organized into the Periodic Table of the Elements.

Groups of Elements

In the periodic table, the elements are arranged in 18 groups. Each column is one group.

Elements in the same group have similar properties. For example, copper, silver, and gold are in group 11. These elements are all shiny metals. They have other properties that are the same, too.

11
29 64 Cu Copper
47 108 Ag Silver
79 197 Au Gold

The Periodic Table of the Elements

6	7	8	9	10	11	12	13	14	15	16	17	18
												2 4 He Helium
							5 11 B Boron	6 12 C Carbon	7 14 N Nitrogen	8 16 O Oxygen	9 19 F Fluorine	10 20 Ne Neon
							13 27 Al Aluminum	14 28 Si Silicon	15 31 P Phosphorus	16 32 S Sulfur	17 35 Cl Chlorine	18 40 Ar Argon
24 52 Cr Chromium	25 55 Mn Manganese	26 56 Fe Iron	27 59 Co Cobalt	28 59 Ni Nickel	29 64 Cu Copper	30 65 Zn Zinc	31 70 Ga Gallium	32 73 Ge Germanium	33 75 As Arsenic	34 79 Se Selenium	35 80 Br Bromine	36 84 Kr Krypton
42 96 Mo Molybdenum	43 98 Tc Technetium	44 101 Ru Ruthenium	45 103 Rh Rhodium	46 106 Pd Palladium	47 108 Ag Silver	48 112 Cd Cadmium	49 115 In Indium	50 119 Sn Tin	51 122 Sb Antimony	52 128 Te Tellurium	53 127 I Iodine	54 131 Xe Xenon
74 184 W Tungsten	75 186 Re Rhenium	76 190 Os Osmium	77 192 Ir Iridium	78 195 Pt Platinum	79 197 Au Gold	80 201 Hg Mercury	81 204 Tl Thallium	82 207 Pb Lead	83 209 Bi Bismuth	84 209 Po Polonium	85 210 At Astatine	86 222 Rn Radon
106 271 Sg Seaborgium	107 272 Bh Bohrium											

59 141 Pr Praseodymium	60 144 Nd Neodymium	61 145 Pm Promethium	62 150 Sm Samarium	63 152 Eu Europium	64 157 Gd Gadolinium	65 159 Tb Terbium	66 163 Dy Dysprosium	67 165 Ho Holmium	68 167 Er Erbium	69 169 Tm Thulium	70 173 Yb Ytterbium	71 175 Lu Lutetium
91 231 Pa Protactinium	92 238 U Uranium	93 237 Np Neptunium	94 244 Pu Plutonium	95 243 Am Americium	96 247 Cm Curium	97 247 Bk Berkelium	98 251 Cf Californium	99 252 Es Einsteinium	100 257 Fm Fermium	101 258 Md Mendelevium	102 259 No Nobelium	103 262 Lr Lawrencium

Numbers and Symbols

The periodic table includes a lot of information about each element.

It shows an element's **atomic symbol**. The atomic symbol for gold is Au. Au comes from *aurum*, the Latin word for gold.

The periodic table also shows an element's **atomic number**. This tells how many protons an atom of the element has in its nucleus. Gold has an atomic number of 79. It has 79 protons in its nucleus.

An element's **atomic weight** is also shown on the periodic table. Atomic weight tells how heavy an element is compared to the lightest element, hydrogen. The atomic weight of gold is 197.

atomic symbol – one or two letters that stand for an element

atomic number – the number of protons one atom of an element has in its nucleus

atomic weight – the weight of an element compared to hydrogen

By the Way...

Dmitri Mendeleev (1834-1907) was the first person to organize the elements by their atomic weight, from lightest to heaviest. He is the "father" of the modern periodic table.

Combining as Compounds

An element is matter made of only one kind of atom. When different elements combine, they form **compounds**. A compound is a molecule that is made of two or more elements.

Water is a compound. A water molecule has one atom of oxygen and two atoms of hydrogen bonded together. Table salt is also a compound. A molecule of salt contains one atom of sodium and one atom of chlorine bonded together.

compounds – molecules that are made up of two or more elements

Compound or Mixture?

When elements bond in a compound, the atoms form a different kind of molecule. So the properties of the compound are different than the properties of the elements that make it up. The elements in the compound cannot be easily separated.

A **mixture** is very different from a compound. A mixture contains different kinds of molecules. The molecules mix together, but the molecules do not change. They keep their properties. They are easy to separate.

Air is a mixture. It contains molecules of several gases. The different gas molecules can be separated from each other easily. Each gas molecule keeps its properties.

mixture – something with molecules of two or more different substances that can be easily separated

Air is a mixture of nitrogen, oxygen, and several other gases.

KEY IDEAS A compound is a molecule made up of two or more elements. A mixture contains molecules of two or more different substances.

YOUR TURN

COMMUNICATE

Choose two elements from the periodic table on pages 10–11. Copy the information about the elements onto two pieces of paper. With a friend, take turns asking and answering these questions about the two elements.

What is the element's symbol?

What is the atomic number of the element?

Which element has a greater atomic weight?

MAKE CONNECTIONS

Many foods are mixtures. Some breakfast cereals are a mixture of nuts, dried fruits, and cereal flakes. Make a list of other foods that are mixtures.

Make Inferences

The elements helium and neon are in the same group on the periodic table. What can you infer about the properties of these two elements? Explain how you can make this inference.

CHAPTER 3

How Matter Changes

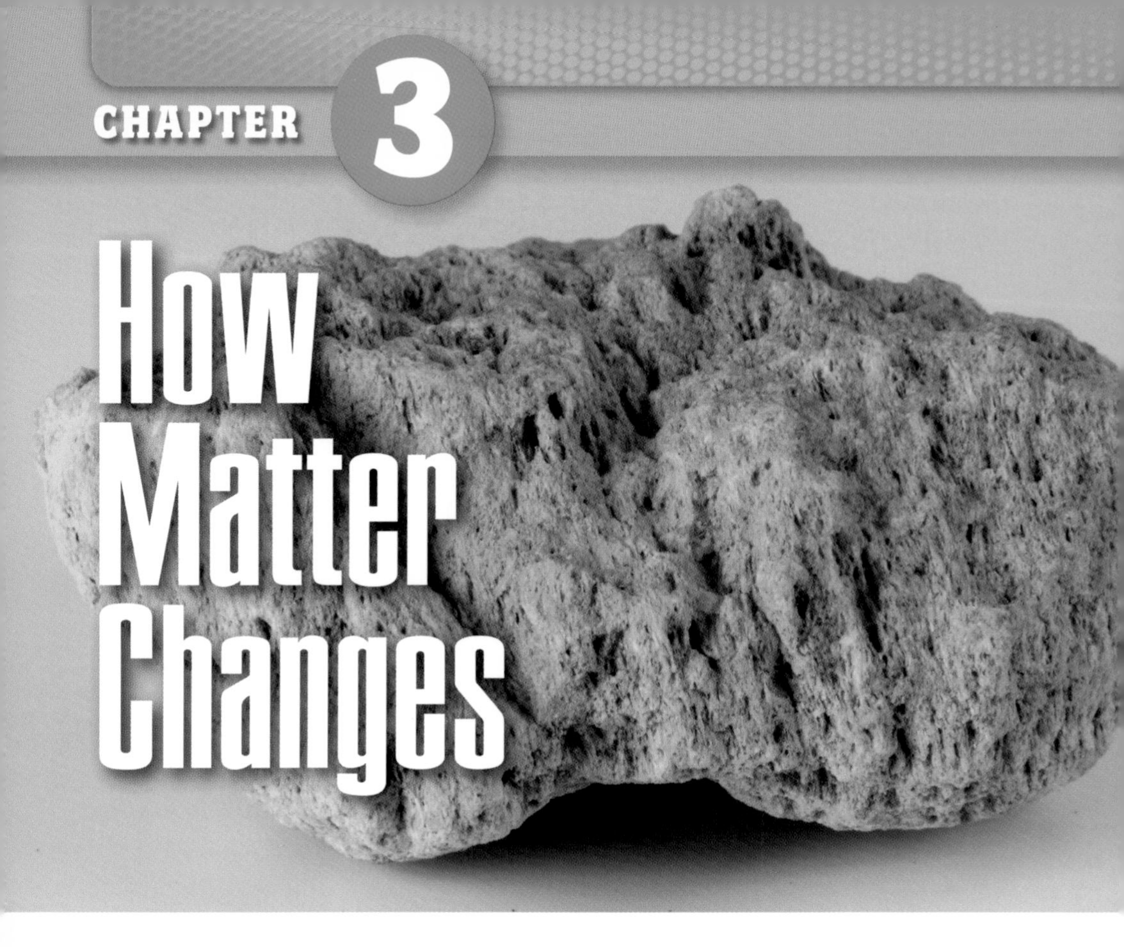

All matter can change. The pumice rock in the picture is matter. If you crushed the rock, it would change shape. It would change into a pile of dust, but it would still be pumice.

▲ Crushing pumice is a physical change.

Crushing rock is a **physical change**. In a physical change, one kind of matter changes in size, shape, or state. It may look different. But it doesn't become a different kind of matter.

physical change – a change in the size, shape, or state of matter

▲ Burning wood is a chemical change that happens quickly.

▲ Rusting metal is a chemical change that happens slowly.

In a **chemical change**, molecules change. One kind of matter turns into a different kind of matter. The properties of the matter are different after a chemical change takes place.

When wood burns, it goes through a chemical change. The properties of the wood change when it is burned. Burning is a chemical change that happens quickly.

When metal rusts, it goes through a chemical change, too. The properties of the rust are different from the metal. Rusting is a chemical change that happens slowly.

chemical change – a change in the molecules that make up a kind of matter

▲ Aluminum foil goes through a physical change when it is crumpled and smoothed.

▲ Cookie dough goes through a chemical change when it is baked.

There is another difference between a physical change and a chemical change. Physical changes can sometimes be **reversed**. Crumpling up a piece of aluminum foil is a physical change. But you can reverse the change. You can unfold the foil and smooth it out again.

Baking cookies is a chemical change. Once the cookies are baked, you can't change them back into dough again.

reversed – changed back

KEY IDEAS Physical changes are changes in the size, shape, or state of matter. Chemical changes produce a new kind of matter.

YOUR TURN

INTERPRET DATA

Pizza is made with dough, cheese, and other toppings. The chart shows what happens to dough and cheese when pizza is made. Interpret the data to answer the questions.

started with	action taken	end result
soft dough	baked in oven	dark and crispy crust
shredded cheese	baked in oven	melted cheese

Which item went through a physical change? Explain.
Which item when through a chemical change? Explain.

MAKE CONNECTIONS

When you eat, your teeth cut and grind food in your mouth. In your small intestine, the molecules of food change. What kind of change is caused by your teeth? What kind of change happens in your small intestine?

EXPAND VOCABULARY

The word *change* can be used in many contexts. For example, you **change** as you grow. But you can also **change** your shirt, **change** your mind, or count your **change**. Write sentences and draw pictures that show different meanings of **change**.

Creating with Clay

Does making pottery out of clay sound fun? Then you might think about a career as a potter. A potter is someone who creates bowls and other objects out of clay. Some potters go to college or art school.

Shaping the clay is the first step. The potter places a lump of clay on a potter's wheel. As the wheel spins, the clay is formed into an object, like a bowl. Shaping the clay is a physical change.

Next, the bowl is painted with glaze and heated. Heating the pot causes chemical changes in the glaze. The glaze forms a strong, hard coating over the clay.

Would you like to be a potter? Tell why or why not.

▲ A potter shapes clay on a potter's wheel.

▲ The glaze on this pot forms a strong, hard covering over the clay.

Use Language to Describe

One way to describe is to tell how something happens.
You can use action verbs to describe a process.

EXAMPLE

Lava **cools** and **hardens** into rock.
A juice bar **melts** on a hot summer day.

With a friend, describe how water changes from solid to liquid to gas. Use action verbs.

Write a Description

Write a description of lava, from when it shoots out of a volcano to when it hardens. Describe how lava moves, how it changes, and how it affects things around it.

- Use action verbs in the present tense.
- Use words that help to create a picture of what you are describing.

Words You Can Use	
flows	cools
shoots	travels
erupts	burns
hardens	

A Popcorn Puzzler

Freshly popped popcorn—yum! It's a crunchy snack people love to eat. But when corn pops, is it a physical or chemical change?

- When popcorn kernels are heated, water inside the kernels gets really hot. The water turns to steam.
- This causes pressure to build up inside the kernels. Eventually, the pressure is so great, the kernels break open.
- The insides of the kernels pop out. They cool and dry instantly to form fluffy popped corn.
- No chemical changes take place. Popping popcorn is a physical change!

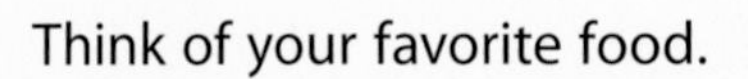

Think of your favorite food.

- Does it go through any physical or chemical changes before you eat it?

Key Words

atom (atoms) the smallest whole unit of matter
An element is made up of one kind of **atom**.

bond join very tightly
Two or more atoms **bond** together to form a molecule.

chemical change (chemical changes) a change in the molecules that make up a kind of matter
Burning wood causes a **chemical change**.

compound (compounds) a molecule that is made up of two or more elements
Salt is a **compound** made of the elements sodium and chlorine.

element (elements) matter made of only one kind of atom
Mercury is a metal **element** that is a liquid at room temperature.

matter anything that takes up space and has mass
Different kinds of **matter** are made of different combinations of atoms.

mixture (mixtures) something with molecules of two or more different substances that can be easily separated
Air is a **mixture** of gases.

molecule (molecules) two or more atoms bonded together
A **molecule** of oxygen is made of two atoms bonded together.

nucleus (nuclei) the central part of an atom
An atom's **nucleus** contains protons and neutrons.

physical change (physical changes) a change in the size, shape, or state of matter
Cutting food is an example of a **physical change**.

property (properties) a quality of matter that can be observed or measured
Chemical changes lead to changes in the **properties** of matter.

state (states) the form matter is in
The matter making up this book is in a solid **state**.

Index

MILLMARK EDUCATION CORPORATION
Ericka Markman, President and CEO; Karen Peratt, VP, Editorial Director; Rachel L. Moir, Director, Operations and Production; Mary Ann Mortellaro, Science Editor; Amy Sarver, Series Editor; Betsy Carpenter, Editor; Guadalupe Lopez, Writer; Kris Hanneman and Pictures Unlimited, Photo Research

PROGRAM AUTHORS
Mary Hawley; Program Author, Instructional Design
Kate Boehm Jerome; Program Author, Science

BOOK DESIGN Steve Curtis Design

CONTENT REVIEWER
Carla C. Johnson, EdD, University of Toledo, Toledo, OH

PROGRAM ADVISORS
Scott K. Baker, PhD, Pacific Institutes for Research, Eugene, OR
Carla C. Johnson, EdD, University of Toledo, Toledo, OH
Donna Ogle, EdD, National-Louis University, Chicago, IL
Betty Ansin Smallwood, PhD, Center for Applied Linguistics, Washington, DC
Gail Thompson, PhD, Claremont Graduate University, Claremont, CA
Emma Violand-Sánchez, EdD, Arlington Public Schools, Arlington, VA (retired)

PHOTO CREDITS Cover © Rob Boudreau/Getty Images; 1 © Gary Cabbe/Alamy; 2, 2-3, 3a, 3b © Patrick J. Endres/Alaska Photo Graphics; 4 © Photo Resource Hawaii/Danita Delimont; 5a © The Natural History Museum/Alamy; 5b and 21 © Jon Arnold Images/Danita Delimont; 6, 7, 10-11 illustrations by Steve Curtis Design; 8a © Peter Witkop/Shutterstock; 8b © William Milner/Shutterstock; 9a, 18a, 18b © Dorling Kindersley/Getty Images; 9b and 9c Lloyd Wolf for Millmark Education; 10 © Photodisc/Punchstock; 12 © Science Source/Photo Researchers, Inc.; 13a © Jose Manuel Gelpi Diaz/Shutterstock; 13b © Bill Beatty/Visuals Unlimited; 14 © Boguslawa Jamka/Shutterstock; 15 © Gabe Palmer/Alamy; 16a © ARCO/O. Diez/age fotostock; 16b Ken Cavanagh for Millmark Education; 17a © Thinkstock/Punchstock; 17b © Jeff Gynane/Shutterstock; 18c © John Wilkes Studio/Corbis; 19 © Dan Peretz/Shutterstock; 20a and 20b © Frank James Fisher; 22a © zimmytws/Shutterstock; 23 © Slobodan Babic/Shutterstock; 24 © Stephen Bonk/Shutterstock

Published by Millmark Education Corporation
7272 Wisconsin Avenue, Suite 300
Bethesda, MD 20814

ISBN-13: 978-1-4334-0058-2
ISBN-10: 1-4334-0058-8

Printed in the USA

10 9 8 7 6 5 4 3 2 1